Cynthia Alsup

Cynthia Alsup

CREATIVE MATH EXPERIENCES FOR THE YOUNG CHILD

Revised Edition

by Imogene Forte and Joy MacKenzie

Illustrated by Mary Hamilton

Incentive Publications, Inc.
Nashville, Tennessee

Library of Congress Catalog Card Number 83-80958

ISBN 0-86530-055-0

Copyright © 1973, 1983 by Incentive Publications, Inc., Nashville, Tennessee. All rights reserved except as here noted.

Permission is granted to the purchaser of one copy of CREATIVE MATH EXPERIENCES FOR THE YOUNG CHILD, Revised Edition to reproduce pages in sufficient quantity to meet the needs of students in one classroom.

CONTENTS

How To Use This Book vii

Learning About Shapes 11
    circle; square; rectangle; triangle

Learning To Read and Write Numbers 33
    numerals and number words, 1-10; zero; counting; sequencing

Counting by Sets 81
    sets; pairs; counting and creating sets

Using Size Words 99
    big, large; little, small; large, small; large, larger, largest; small, smaller, smallest; short, long; long, longer, longest; short, shorter, shortest; tall, taller, tallest; more, fewer; greater than, less than

Finding Parts of Things 121
    wholes; halves

Living With Numbers 129
    numbers in everyday life; addresses; telephone numbers; money; numbers in games and races; weight; linear measure; numbers for finding things; directories; menus; tables; calendars; clocks; liquid and dry measure; temperature; numbers for travel; gauges; meters; computers

Test Yourself 168

Certificate of Achievement 175

HOW TO USE THIS BOOK
(Instructions for parents and teachers)

Many creative teachers have found that children can deal effectively with mathematical concepts at an earlier age than has been previously thought and that they can and will discover mathematics for themselves if appropriate situations are provided. The young child's environment should be saturated with opportunities to use numbers in natural and meaningful settings. As children are guided in counting objects and people; in learning the use of numerals on clocks, calendars, measuring cups, rulers and scales; and in observing the variations and relationships in sizes and shapes of objects, they are building beginning mathematical concepts. It is important to remember that they learn on an individual basis and at their own rate, and that it is in this context that mathematical experiences should be presented.

The mastery of these all-important foundational concepts and understandings is developmental in nature and should be presented to children in proper sequence at a strategic time in keeping with their readiness in terms of their past experiences and presently recognized needs. This phase of the young child's intellectual development should never be left to chance. It should be planned and carefully monitored by an adult who is sensitive to the rate of progress being made and to the need for redirection or reinforcement in specific areas. It is important to remember that learning is continuous and developmental, and that each stage of mathematical growth is dependent upon success in the stage that preceded it. In keeping with this belief, the activities in *Creative Math Experiences for the Young Child, Revised Edition* are sequentially planned and should be introduced to the child as they are presented in the book.

To provide help in determining when a child has mastered the basic concepts, skills and factual information related to each of the six areas in *Creative Math Experiences for the Young Child, Revised Edition,* a review is included at the end of each section. The review should be administered and scored with adult guidance, and immediate feedback should be furnished the child.

Learning About Shapes

As young children are exposed to different shapes, they learn to recognize the shapes and to identify them in their natural environment. Transfer of this awareness to patterns in sidewalks, clothing, kites, buildings and other everyday objects heightens sensitivity to the world. As they add the words *triangle, rectangle, square* and *circle* to their vocabularies, important concepts related to skill development in both language and math are being established. Matching, copying and reproducing these shapes helps the child acquire the all-important visual discrimination skills which are prerequisite to reading readiness.

These skills are also directly related to perceptual development. Perception simply means the ability to transmit stimuli to the brain and to interpret them accurately. While all the five senses are included under the perceptual tag, visual discrimination is the key to early learning. What children can see they can understand. Research indicates that up to fifteen percent of the children in the average first grade class are hindered in the development of good reading skills by some form of perceptual difficulty.

Each of the activities in this section is designed to help children fulfill these basic developmental needs. Adult guidance in completing the activities should be consistently patient, yet firm, and should be adjusted to the child's natural rate of progress. If extended training appears to be in order, additional worksheets using the same basic format as the ones completed can be designed.

Learning To Read and Write Numbers

Counting by rote is often mistaken as evidence of a child's understanding of the counting process or of readiness to work with abstract number concepts. Only when he/she is able to give meaning to rote counting is it of consequence. Real objects that can be touched as they are counted should be included as a regular part of early counting experiences. Use of rhyming words, finger plays and games will also enrich and strengthen the understanding being gained. Figures should always be called by name and not referred to as numbers. These activities have been planned to give experience in meaningful counting and to extend concepts to include use of printed symbols and related vocabulary. While it is hoped that readiness for writing numerals will be a natural by-product of these activities, lack of pressure cannot be too strongly emphasized.

Counting by Sets

Children need to understand that sets are simply groups or collections of objects. By grouping things into sets and discovering how sets may be manipulated, concepts of union and subsets are developed. It is through counting objects in various sets that the cardinal meaning of numerals can be developed. By developing matching sets of the same value, the child learns the meaning of *three-part* or *seven-part*. Number language will naturally emerge from meaningful experiences with sets of numbers. More than, less than, smaller than, larger than, and other mathematical relationships become well understood and useful additions to the child's intellectual assets.

If further exposure to activities of this nature appears to be in order, additional worksheets can be modeled on the completed ones. Care should be taken to duplicate the format and style of activities in *Creative Math Experiences for the Young Child, Revised Edition,* but to use new and different combinations and illustrations. The boredom which comes from recognizable repetition is to be avoided at all costs. Young children respond positively to the excitement and the challenge of the "first time" experience.

Using Size Words

Remembering the importance of concrete experiences to the attainment of meaningful concepts, size words and other vocabulary extensions must be planned as an integral part of the sequentially developed math program. The child's encounter with them should be more than incidental. Understanding and use of these words enhance the quest for in-depth and more abstract learning. Their presentation will provide the base for the emergence of number language in the most meaningful sense. Completion of these worksheets with adult guidance adjusted to the child's individual needs should provide the reinforcement necessary to solidify understanding of size words and their application to problem-solving situations.

Finding Parts of Things

Use of fractions should be limited to basic beginning concepts and in most cases should not be extended beyond the *whole* and *half* stage. Simple use of parts of a whole and the acquisition of related vocabulary is the chief concern of the activities in this section. Verbalization of terms such as *both, divide, double, middle, alike,* and *two parts* is desirable as the activities are being completed. Reinforcement can also be afforded through directions such as "fold your sheet of paper in the middle and use only one half of it for your picture" or "divide the stack of pennies in half."

Under no circumstances should children be exposed to new situations related to fractions and their use before these basic concepts are well established. If additional experiences are in order, they should be planned to center on the use of concrete objects and to afford lots of opportunity for sensory experiences such as cutting an apple in half so that it can be shared by two people. Remember, there is nothing to be gained from "teaching" children skills they are not ready to use. On the other hand, they must not be held back by lack of experiential exposure from pursuing their own intellectual curiosity. The readiness for fractions and their use is as sensitive an area as any in the beginning math picture, and must be judged on an individual basis. One should proceed with caution and guard against being overly ambitious.

Living With Numbers

Very little learning about numbers is worthwhile if there is no application to real-life situations. Does the child seem to recognize the importance of numbers to everyday existence? Is he/she aware of the use of numbers as they appear in elevators, on buses, TV sets, menus, telephones, mailboxes and license plates?

In assessing individual readiness for the development of concepts of measurement, it is important to listen to the child talk to determine if his/her vocabulary includes words related to shape, size, amount, capacity, time and distance. It is equally important to observe the awareness of objects of measurement in his/her daily environment. The child whose attention has been called to the clock and the calendar as important tools for social living, to the thermometer and scales as useful for gaining desirable information, and to measuring cups and spoons as indispensable in the kitchen is much more apt to approach the worksheets in this section with zeal and commitment.

Today's children are naturally curious about and very much interested in money. They know from an early age that certain objects can be had or not had because of money. From television commercials and the conversation they hear, they know that the adults around them are aware of money. As activities in this section are presented to the child, discussion of cost of toys or sweets, items in catalogs and the money in the piggy bank will be extremely beneficial. Trips to the store and an opportunity to spend a small amount of money as he/she sees fit will also serve to give meaning to developing concepts related to money.

There is no substitute for many and varied real experiences which are planned in sequence to be developmental in nature. For most children, the activities in *Creative Math Experiences for the Young Child, Revised Edition* will be motivational in nature, and should be solidly reinforced by extensive discussion, trips, games and manipulative experiences.

LEARNING ABOUT SHAPES

Can you name these things?

Can you tell how all of them are alike?

If you said they are all round, you are exactly right.
Lines that are round in shape are called <u>circles</u>.
Trace the round circle shapes with your finger.

Color each circle a happy color.

©1983 Incentive Publications, Inc., Nashville, TN. All rights reserved.

How many circles can you find in the picture?
Make a green mark on each circle.

Finish these pictures by adding some circles.

Use each group of circles to make a picture of your own.

©1983 Incentive Publications, Inc., Nashville, TN. All rights reserved.

15

This shape is a <u>square</u>.
How many sides does it have?
_____

How many corners does it have?
_____

If you said four, you are exactly right.
The sides of a square are all the same length.
All of its corners are the same.

Trace the squares with your finger.
Color each square orange.

Can you find the square shapes in the pictures below? Trace them with your finger.

Color the pictures you like best.

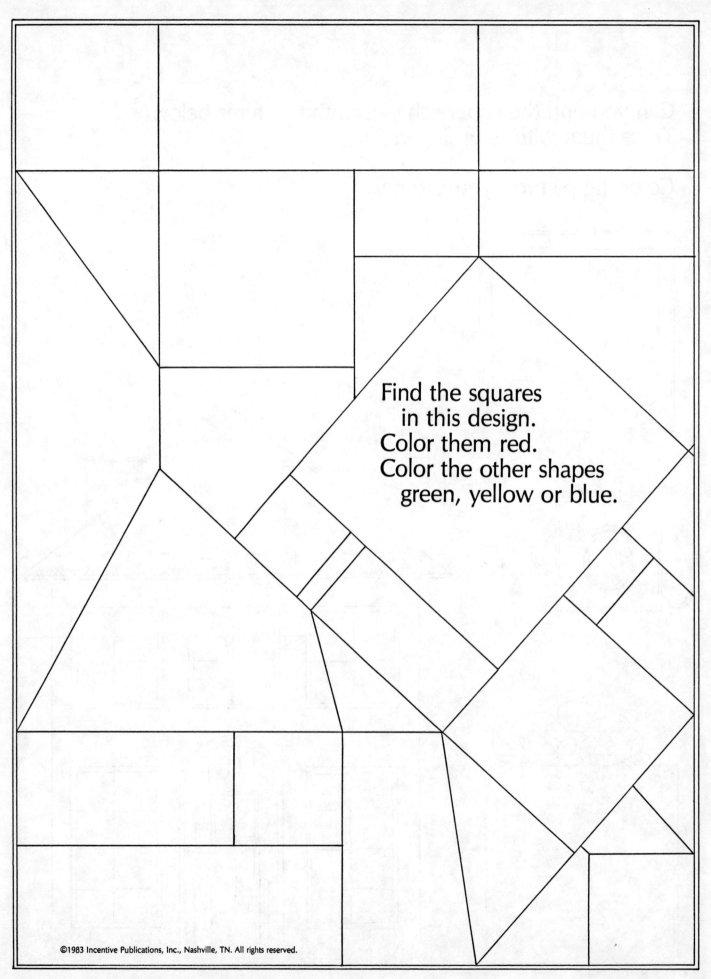

Build a House

Color all these squares.
Then carefully cut them out along the dotted lines and paste them on the squares of the house.

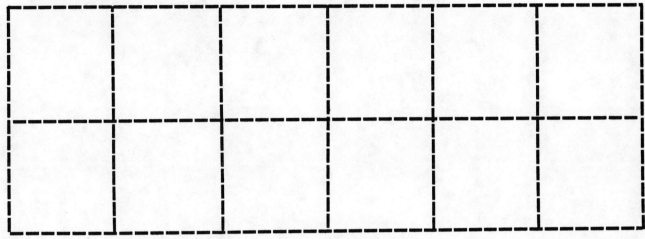

See what a lovely house you can make.

Do you know what these shapes are called?
They are <u>rectangles</u>.
That's a big word. Say it three times.
RECTANGLE!
RECTANGLE!
RECTANGLE!
Rectangles have corners like squares, but their sides are not always the same length.
Only the sides that are opposite each other are the same length.
Color the rectangles.

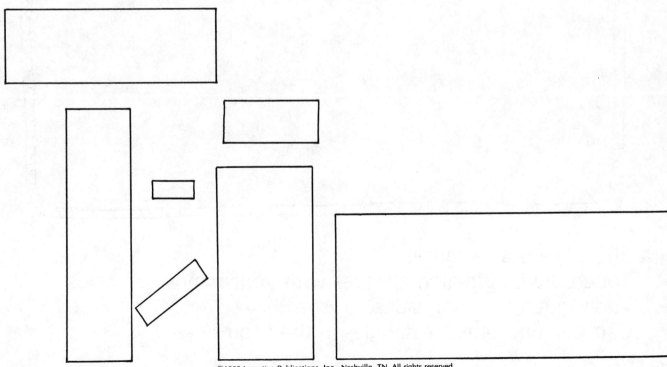

A rectangle that is not a square has two long sides that are the same, and two short sides that are the same.

Look at these rectangles. Put your finger on their long sides.

Now put your finger on their short sides.
Trace the big rectangle with your finger.

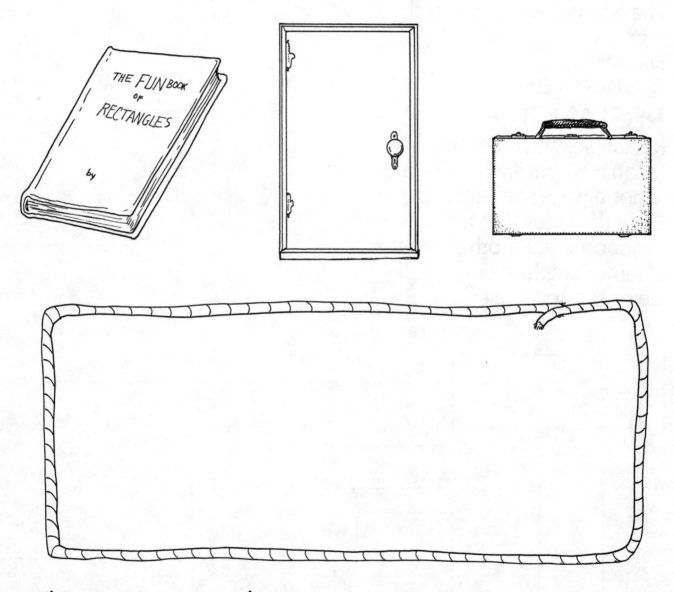

This page is a rectangle.
Touch the long matching sides with your hands.
Then touch the short sides.
Can you find other rectangles in the room?

Rectangle Roundup

Find a big, empty space on the floor or go outside on the sidewalk or grass.
Use a long piece of string to make a big rectangle.

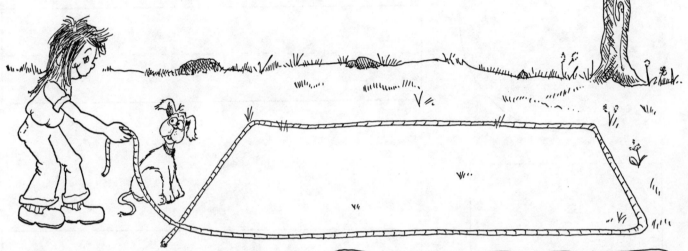

Walk around your rectangle.
Hop around your rectangle.
Run around your rectangle.

Lie down on its long side.
Ask a friend to lie down on its other long side.
Then you and your friend stand opposite one another on its short sides.
Wave to each other.

Find a small object that is the same shape as each figure below.
Place each object on the matching shape and trace around it.

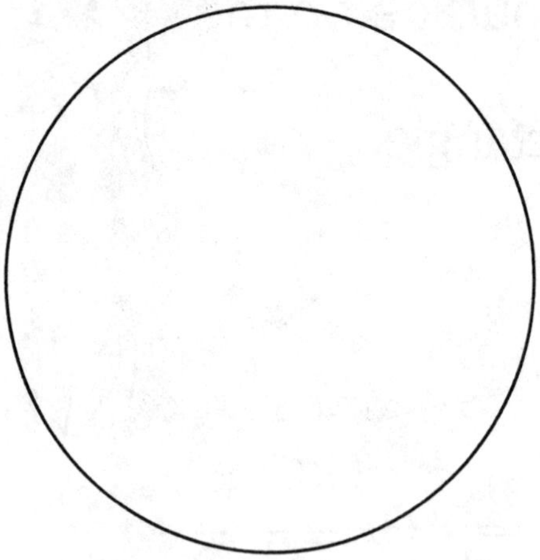

Write the name of each shape on the line.

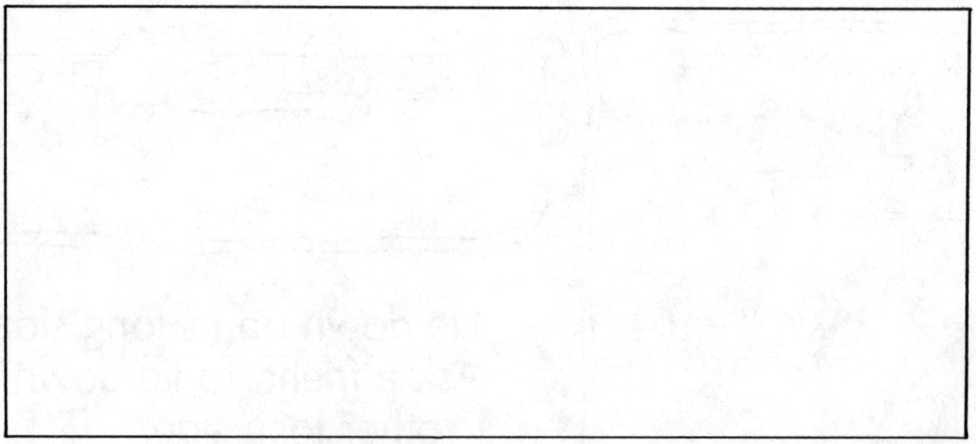

24

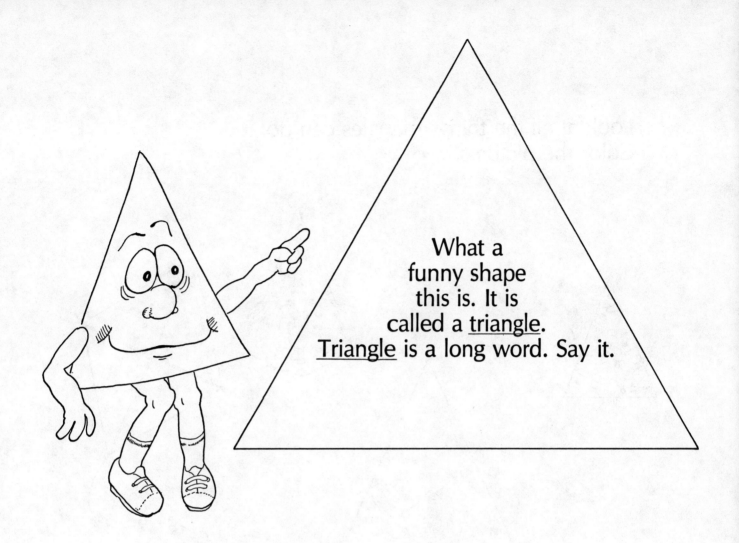

What a funny shape this is. It is called a <u>triangle</u>. <u>Triangle</u> is a long word. Say it.

A triangle has three corners.
A triangle has three sides.
Trace the triangles with your finger.
Color them yellow.

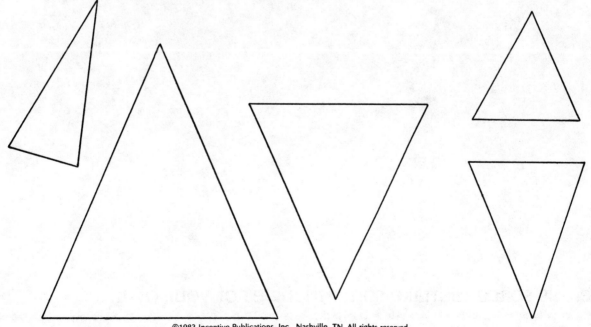

Look at all the things triangles can do!
Color the pictures.

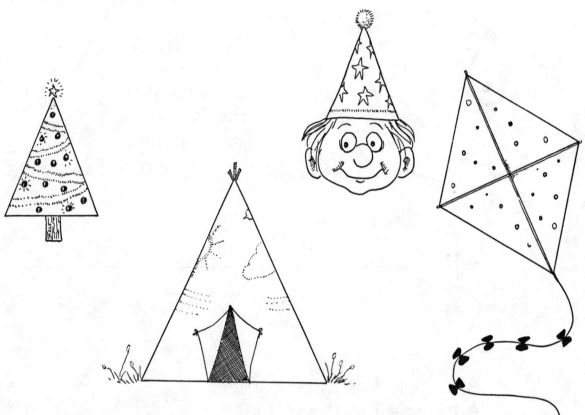

Use this space to make some triangles of your own.

A Triangle Trick

Carefully cut out these triangles.

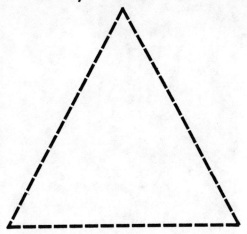
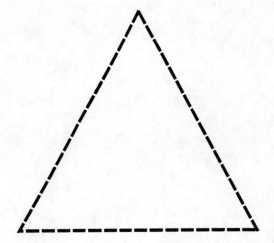

Paste them together like this.

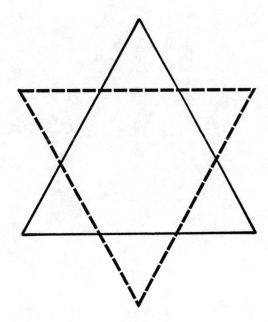

What new shape did you make?

Clown Cut-Up

Use your crayons to color these shapes.

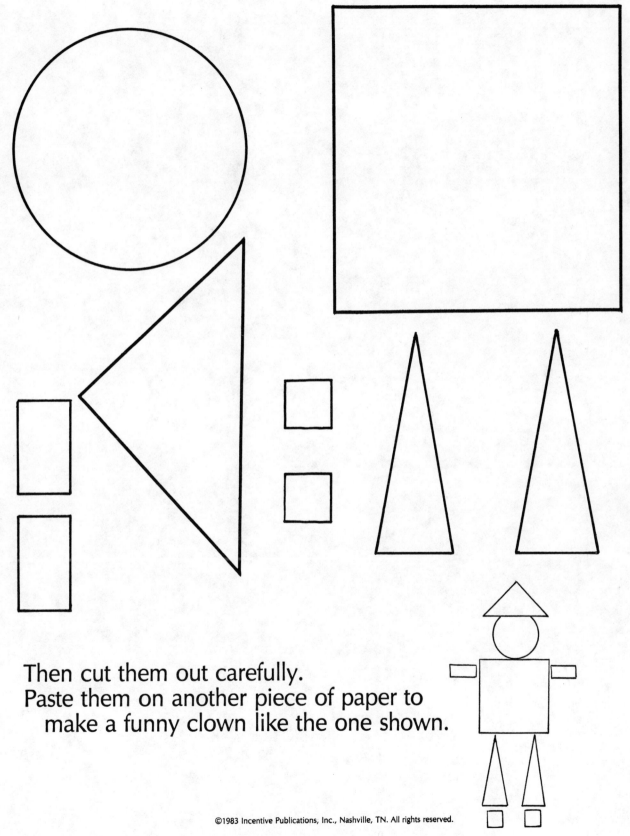

Then cut them out carefully.
Paste them on another piece of paper to make a funny clown like the one shown.

©1983 Incentive Publications, Inc., Nashville, TN. All rights reserved.

LEARNING TO READ AND WRITE NUMBERS

How many hats are in this picture?
Color the picture pretty colors.

Practice writing the numeral.

One little girl is playing grown-up.

Write the number word.

How many scarecrows are here?
Color the scarecrows.

Practice writing the numeral.

Count the jack-o'-lanterns.
How many are happy?

Write the number word.

How many mice are here?
Color one mouse yellow.
Color two mice brown.

Practice writing the numeral.

How many cakes are here?
Color the cake that the mice have been eating.

Write the number word.

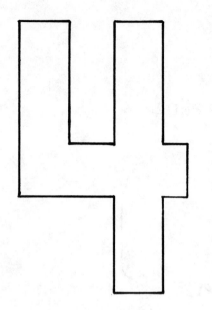

How many turtles are here?
Color the turtles' hats.

Practice writing the numeral.

Count the rabbits.
They are racing with the turtles.
Which do you think will win?
Pick a winner and color it gray.
Color the others green.

four

Write the number word.

Color the frogs.
Use your red crayon to color a tie on the one who is singing.

Practice writing the numeral.

five

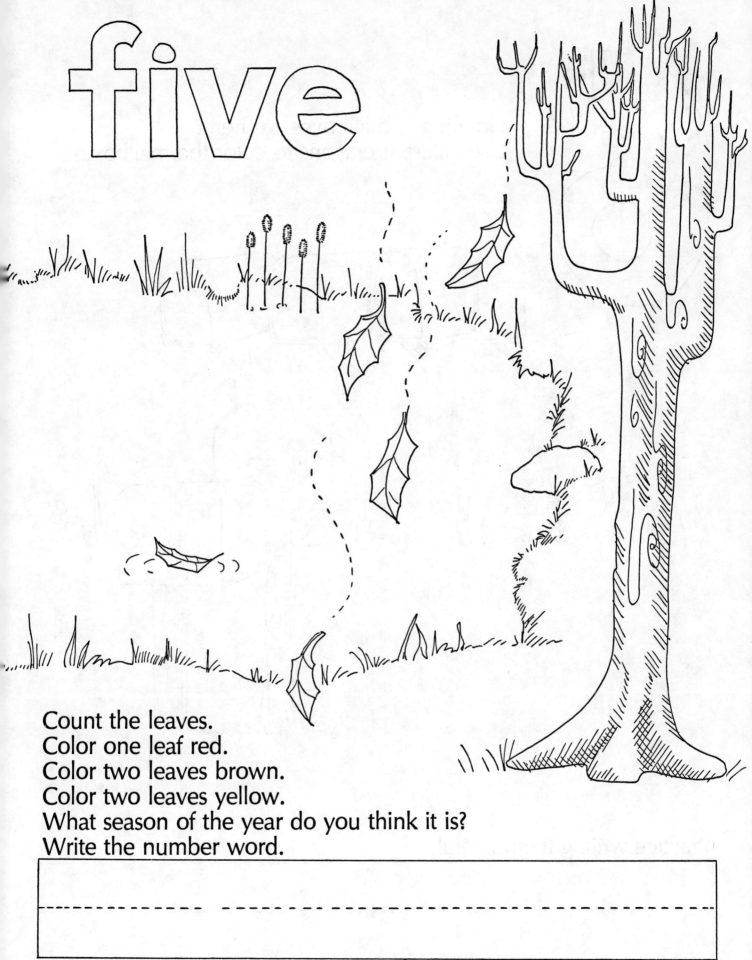

Count the leaves.
Color one leaf red.
Color two leaves brown.
Color two leaves yellow.
What season of the year do you think it is?
Write the number word.

6

How many mailboxes are here?
Use different crayons to color the mailboxes.

Practice writing the numeral.

How many letters are here?
Color one letter yellow.
Color two letters blue.
Color three letters red.
Draw six letters in the mailbag.

Write the number word.

How many apple trees are in the picture?
Count and color them.

Practice writing the numeral.

Count the baskets.
Put an apple in each basket.
Count the apples.

Write the number word.

8

How many flowers are here?
Color the flowers many beautiful colors.

Practice writing the numeral.

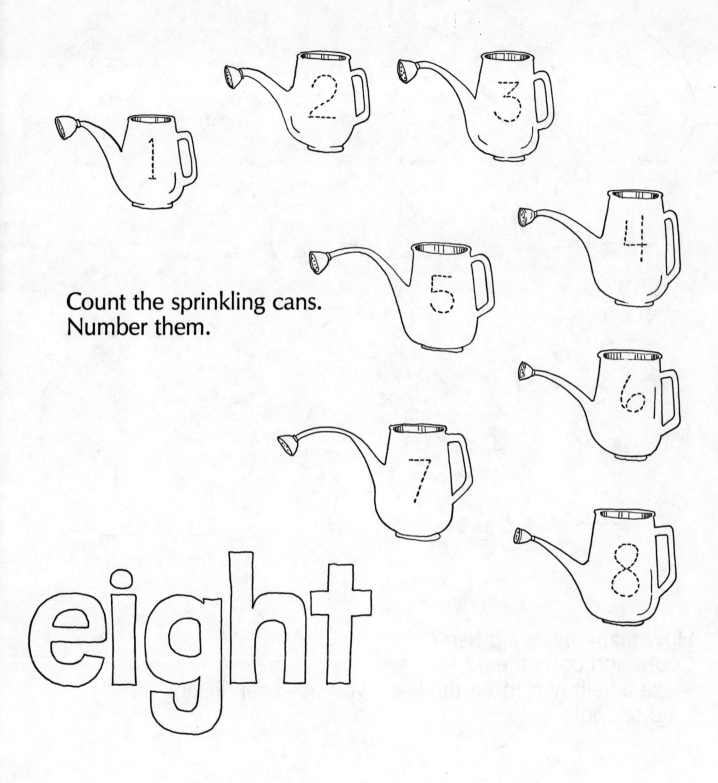

Count the sprinkling cans.
Number them.

eight

Write the number word.

How many bears are here?
Count and color them.
Make a yellow chin on the bear who has been licking the honey pot.

Practice writing the numeral.

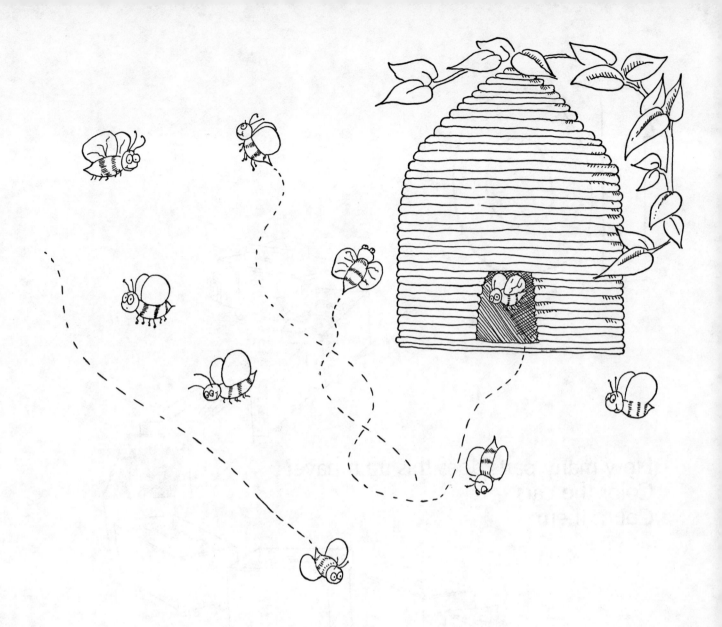

How many honeybees are here?
Color the bees yellow.

Write the number word.

10

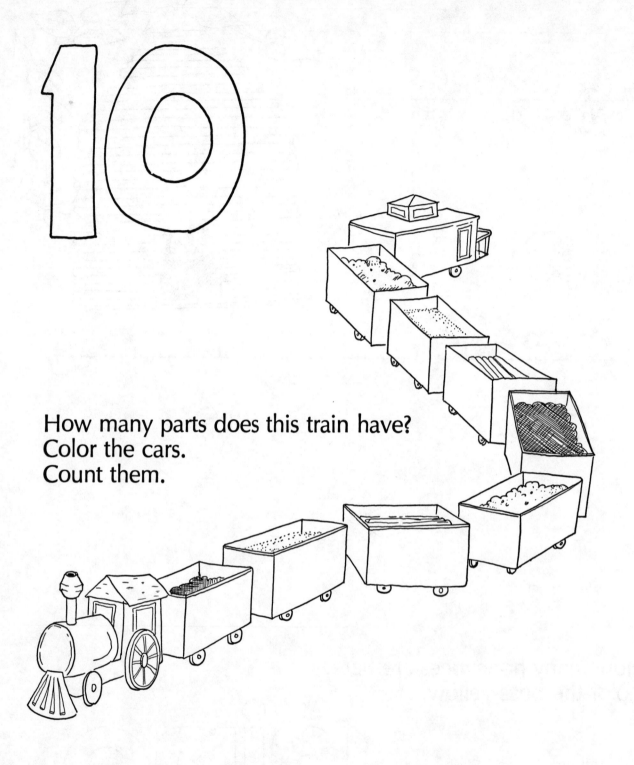

How many parts does this train have?
Color the cars.
Count them.

Practice writing the numeral.

Write the numerals 1-10 in the windows of this station house.

Write the number word.

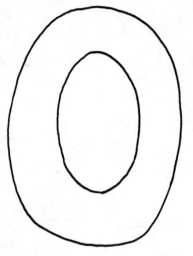

This number is zero.
Zero means having none.
Make a zero in this box.

A Story About Zero

Betty had two kites.

The wind blew both of them away.

Now Betty has no kites at all. She has none.

Make a zero to tell how many kites Betty has.

©1983 Incentive Publications, Inc., Nashville, TN. All rights reserved.

How High?

Count the figures in each stack.

How many?   How many?   How many?

Which stack has the most? Circle it.

How Many?

Write the numerals.

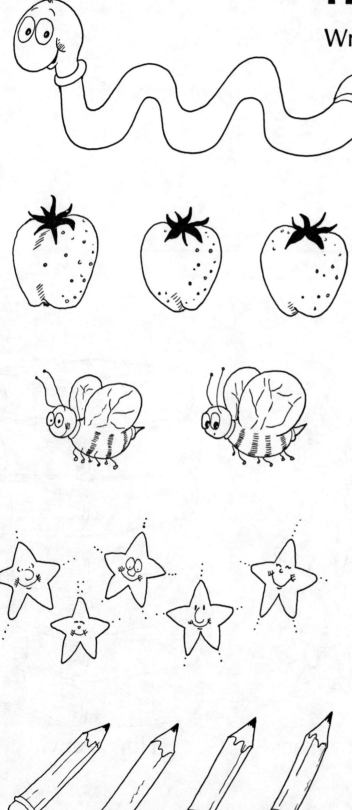

_____

_____

_____

_____

_____

Write each number word.

1

2

3

4

5

Feed the Rabbit

Trace the numeral in each patch. Use your orange crayon to color that many carrots in each patch.

Fish 'N Fun

Trace the number.
Then use your orange crayon to color that number of fish in each bowl.

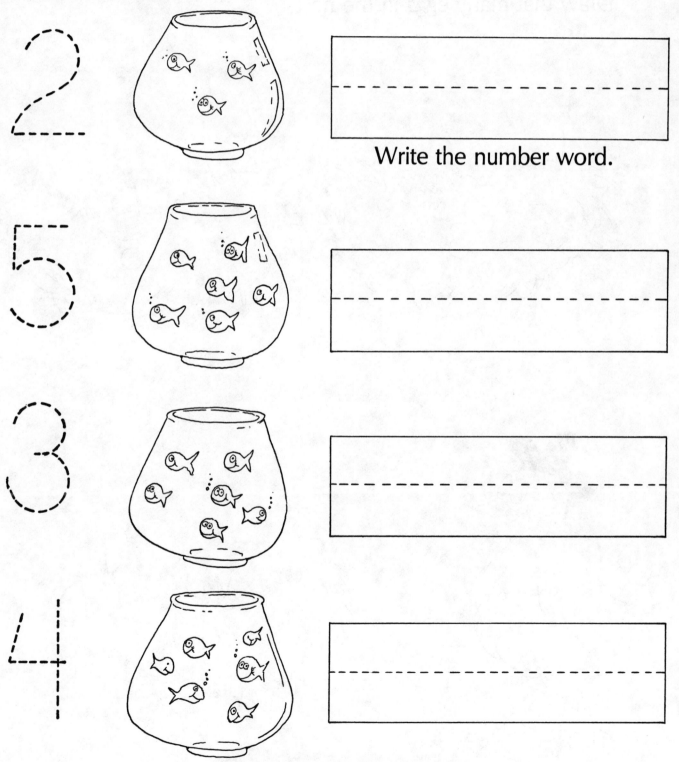

Write the number word.

Chicken Coop

Read the numeral on each nest.
Draw that many eggs in the nest.

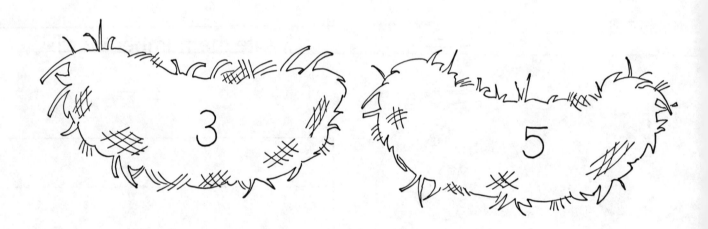

Jumbo Gum-bo!

Count the jawbreakers in each gumball machine.
Write the numeral in the box.
Write the number word on the line.
Use bright colors to color the jawbreakers.

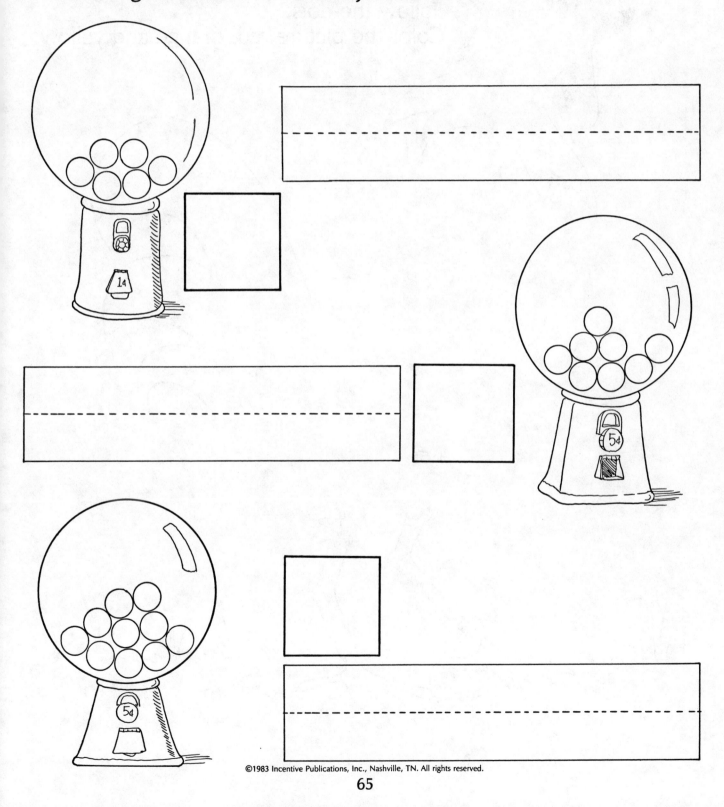

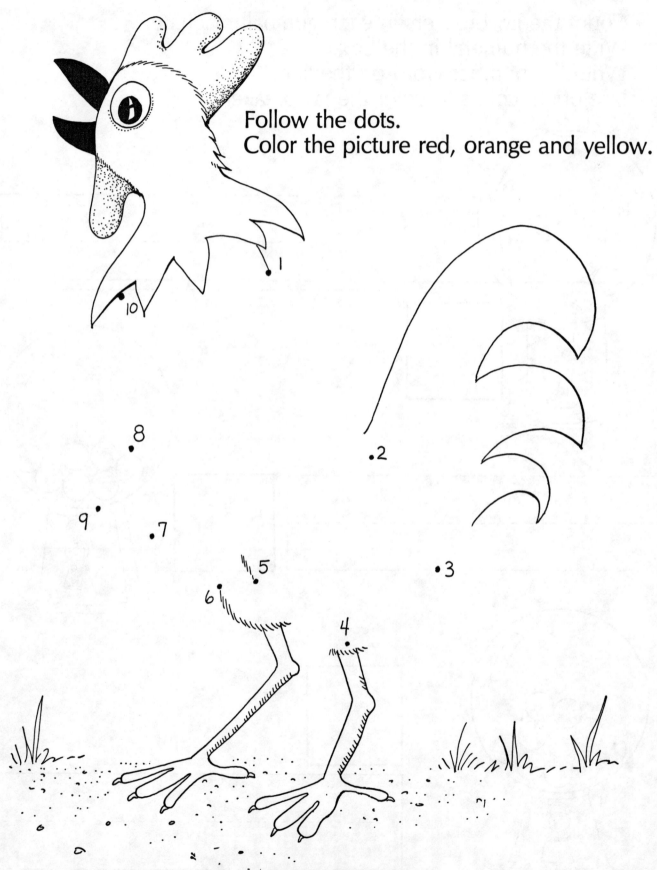

Find all the 5's on this page.
Circle each 5.
How many 5's did you find? _____
Can you point to each numeral and say its name?

Shutdown for Robot Repair!

The robot's keyboard has self-destructed.
Help repair it by writing the missing numerals.

1	2	3	4	5
3	4	5	6	
	7		9	10
2		4		
	5		7	
	2		4	
5	6			9
4		6		

Follow the dots to catch a whopper.
Color it with pretty colors.

Who's in the Barnyard?

Count to find out how many of each animal are in the barnyard.
Write the correct numeral in each box.

| ☐ |  | ☐ | | ☐ | |

This hungry worm has eaten some of the numerals. Can you supply the missing numerals?

Climb the Tipsy Tower!

Number the parts of the tipsy tower from bottom to top.

Which numerals can you find in this puzzle?
How many are there?
Color them yellow.

All Tied Up!

Finish tying together this bunch of balloons. Then number each balloon to find out how many there are.

Finish stringing the beads. Then number each bead to find out how many there are.

Number Puzzles

Finish the crossword puzzles by completing the number words.

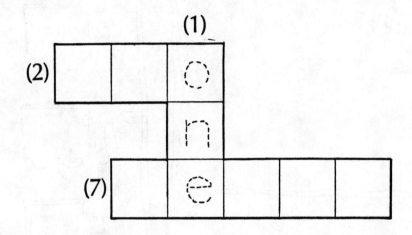
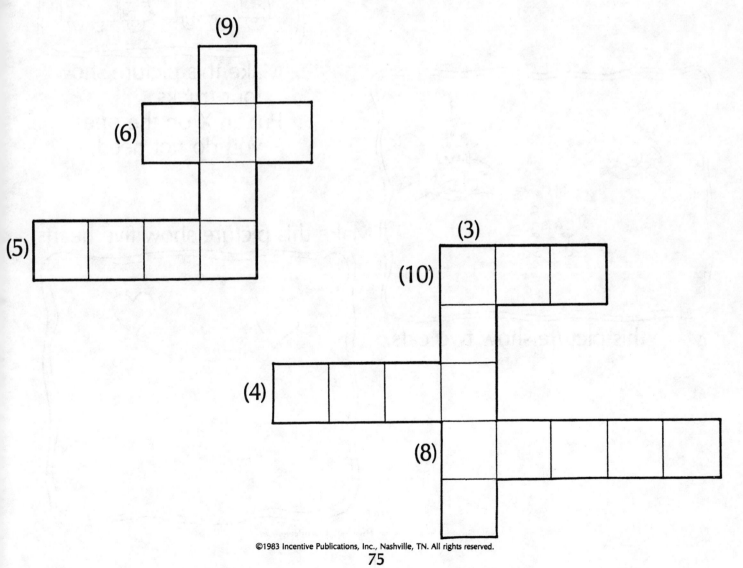

Picture Talk

Add hats to make this picture show six.

Make this picture show four trucks.
Put an X on the ones you do not need.

Make this picture show two cats.

Make this picture show five hearts.

Make Your Own Book

Cut on the solid lines.
Fold on the dotted lines.

Use your crayons to make each page of your book beautiful.

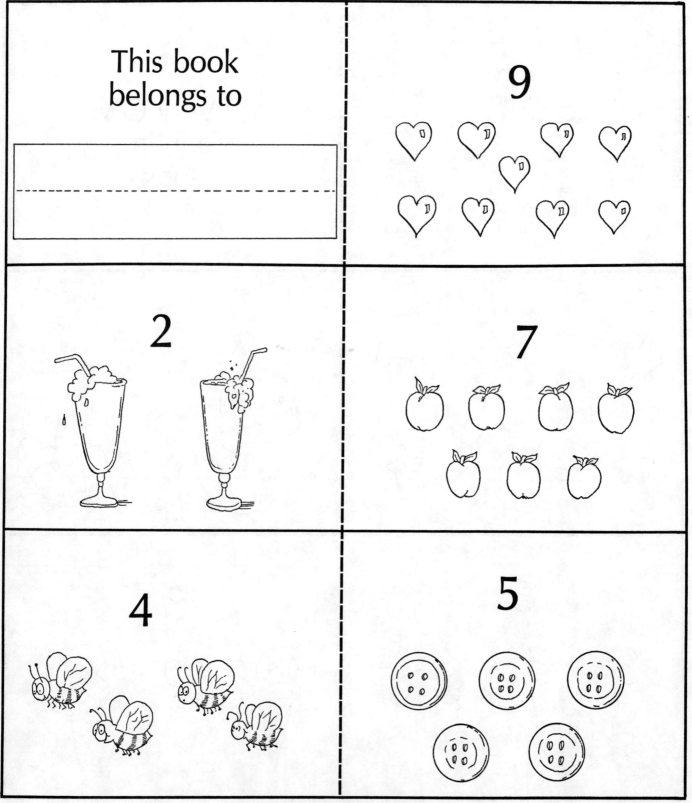

What Have You Learned?

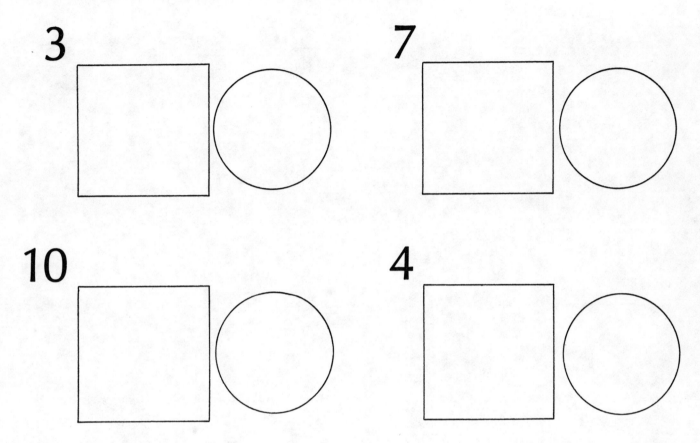

Cut on the dotted lines.
Match the words and pictures with the numerals above and paste each piece in the proper place.

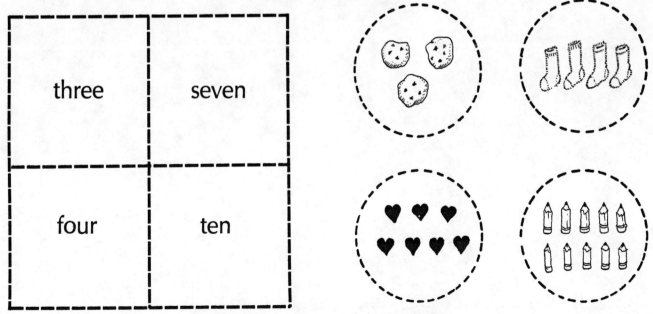

©1983 INCENTIVE PUBLICATIONS, Inc., Nashville, TN. All rights reserved.

COUNTING BY SETS

A <u>set</u> is a group of things that are alike and together.

This is a set of three owls.
How many owls are in the set?
Write the numeral.

_____

This is a set of two umbrellas.
How many umbrellas are in the set?
Write the numeral.

_____

How many teapots are in this set?
Count them.
Write the numeral.

_____

Each group of sweets is called a set of sweets.

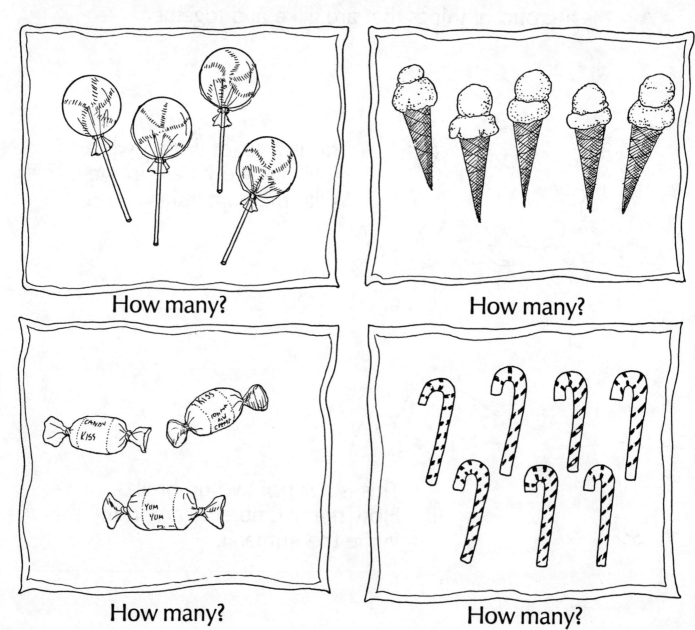

How many?

How many?

How many?

How many?

How many different sets did you find? _____

Who's Hiding Under the Stairs?

Who's hiding under a set of 3 steps?

Who's hiding under a set of 4 steps?

Who's hiding under a set of 2 steps?

Who's hiding under a set of 5 steps?

©1983 Incentive Publications, Inc., Nashville, TN. All rights reserved.

Read each number below.
Write the correct numeral in each square.
Then draw a set to match that numeral in the rectangle.

two   2

four

six

three

five

How many are in each set?
Write the numeral.

Spider Sets

Color the set of two spiders green.
Color the set of four spiders orange.
Color the set of three spiders blue.

Pairs

This is one shoe.

This is one pair of shoes.

A pair is a set of two things that go together.

Draw one shoe for a giant.

Draw a pair of mittens.	Draw a pair of socks.

Can you think of other things that come in pairs?

Sets of Pairs

Match each numeral with a set of pairs.

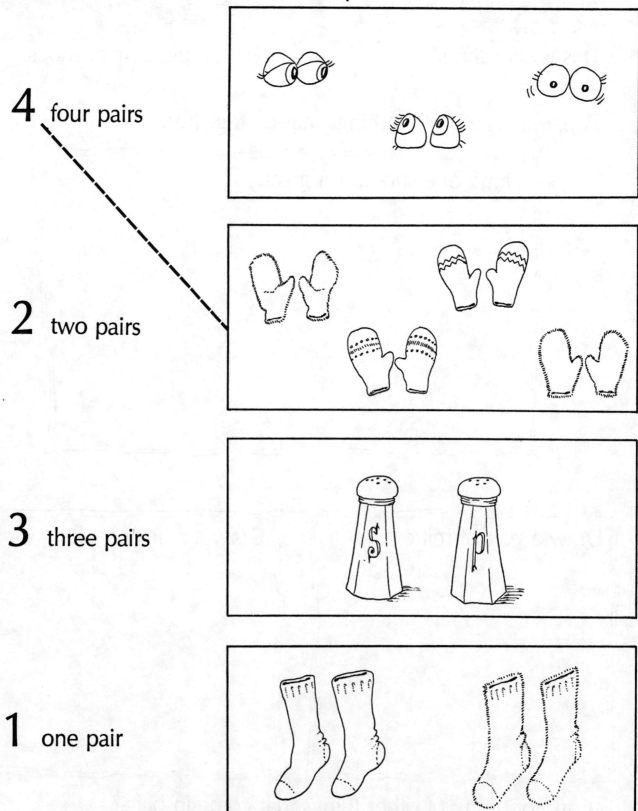

4 four pairs

2 two pairs

3 three pairs

1 one pair

Here, Spot!

Find three pairs of dogs who have matching sets of spots. Give each set of twins matching collars by making one pair red, one pair blue, and one pair green.

Bee Busy!

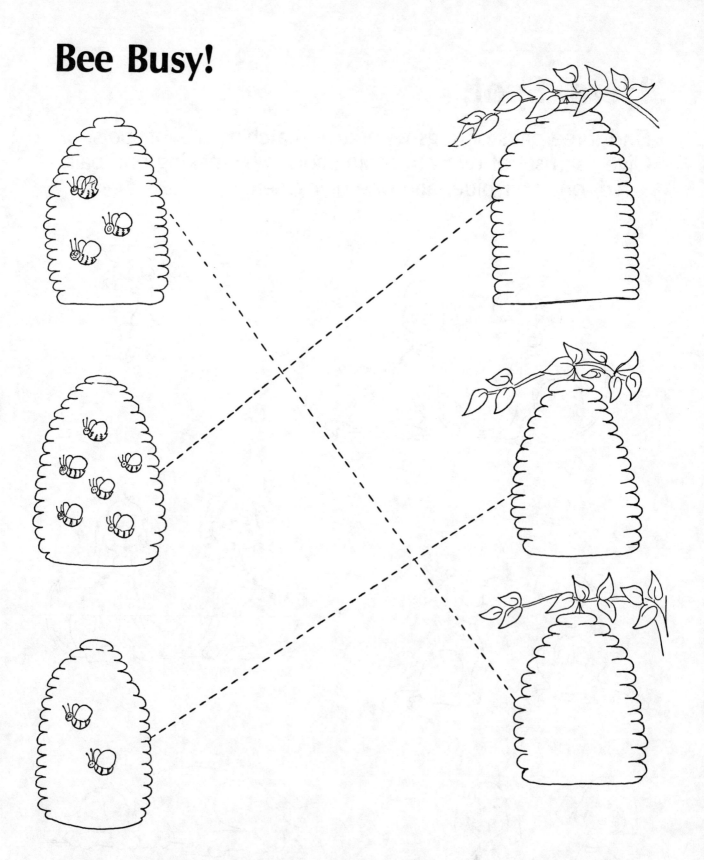

Draw a set of bees in each empty hive to match the set in each connecting hive.

Boxes 'N Bows

If the set on each package matches the number word below it, put a bow on the package.
If it does not match, put an X on the package.

four

five

six

eight

seven

three

Dinner's On!

To do this page, you can use ten real pieces of corn or cereal, or you can draw the pieces.

Feed the puppy a set of two pieces.

Feed the chick a set of four pieces.

Feed the pig all the rest.
How many pieces did the pig get?

Connect-a-Set

Make triangles to connect each of the three matching sets. What new shape did you make?

Creature Feature

Complete the creature by drawing these sets on its body.

a set of two ears
a set of six hairs
a set of five teeth

a pair of hands
one big nose

USING SIZE WORDS

Big

These are pictures of big things.
Another word for big is <u>large</u>.

Color the big tree green.
Color the large bear brown.
Color the big airplane blue.
Color the large barn red.

Large

An elephant is a very large animal.
Write the word <u>large</u>.

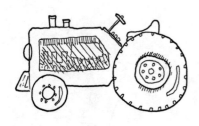

A tractor is a big machine.
Write the word <u>big</u>.

Use this space to draw something that is big or large.

Little

These pictures show things that are little.
Another word for little is <u>small</u>.

Draw a circle around the little bird.
Make a box around the small bug.
Can you put the little fish inside a big bubble?
Color all the other small things.

Small

See the tiny little mouse.
Write the word <u>little</u>.

This baby chicken is small.
Write the word <u>small</u>.

See how many little or small things you can draw in the space below.

This face is large.

This face is larger.

This face is the largest.

Make a red nose on the largest face.
Color the eyes on the smallest face blue.
Make orange hair around the face that is neither the largest nor the smallest.

This mouse is small.

This mouse is smaller.

This mouse is the smallest of all.
Color the smallest mouse yellow.

 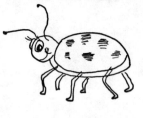

Which bug is the smallest? Give it a leaf to sit on.

Which feather is the smallest? Color it orange and red.

Scratch the largest of each group with your fingernail.
Then put a big X on it.
Make a circle with your crayon around the smallest of each
   group.

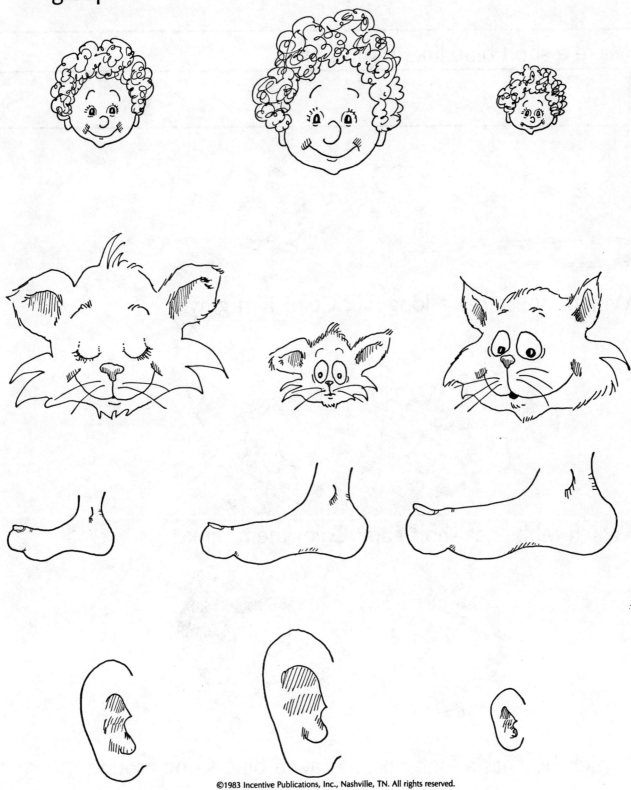

The Long and Short of It

Make a long red line in this box.

Make a short blue line in this box.

Which mouse has a long tail? Color him gray.

Which rabbit has short ears? Color them pink.

Which boy has a long nose? Draw a blue string around it.

Tall, Taller, Tallest

This ladder is tall.   This ladder is taller.   This ladder is tallest.

Make your fingers walk up the tallest ladder and then write your name at the very top.

Make the shortest ladder reach a shelf with a cookie jar on it.

Use your yellow crayon to color the ladder that is neither the tallest nor the shortest.

The road to the lake is long.
The road to the bridge is longer.
The road to the airport is longest.

Trace the longest road with
  your longest finger.
Make flowers along the shortest road.
Put a car on the road that is neither the longest nor the
  shortest.

The road to the church is short.
The road to the store is shorter.
The road to the woods is the shortest of all.

Color the shortest road brown.
Make a purple line on the longest road.
Color grass along the road to the store.

More/Fewer

The big house has ten windows.
The little house has four windows.

The big house has more windows than the little house.
The little house has fewer windows than the big house.

Which cookie has more chips? How many chips does it have?

_____

Which cookie has fewer chips? How many chips does it have?

_____

Which hat has more feathers? Color that hat blue.

Which plate has fewer cookies? Color that plate yellow.

Which flower has fewer petals? Color it a pretty pink.

Which bunch has more balloons? Color it.

Greater Than/Less Than

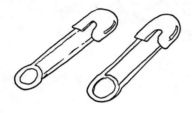 is greater than

 is less than

 is greater than

Which is greater?   Circle the greater number in red.

   or

Which is less?   Put an X on it.

   or

Size Words for a Super-Smart Kid

Fill the smallest jar with purple jelly.

Use your yellow crayon to light the tallest candle.

Put an X on the larger pair.

Catch the biggest bug in a circle.

Give the tallest flower leaves.

Give the smallest butterfly a leaf to rest on.

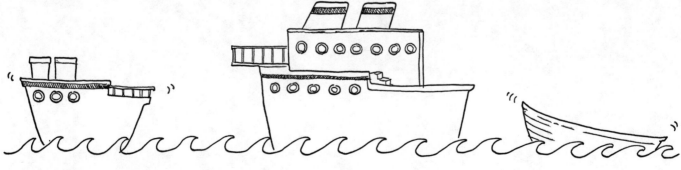

Make a sail for the shortest ship.

©1983 Incentive Publications, Inc., Nashville, TN. All rights reserved.

FINDING PARTS OF THINGS

Whole/Half

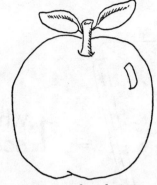

This is one whole apple.

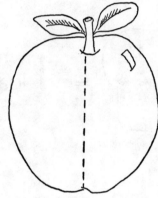

Make a line to cut the apple in half.

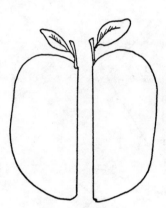

Now there are two halves. Each part is one half.

We write one half like this:
Write ½ on each half below:

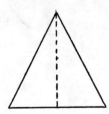 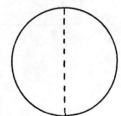 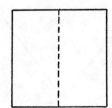 

When anything is divided in half, both halves must be exactly the same size.

When two halves of something are put together, they make one whole.

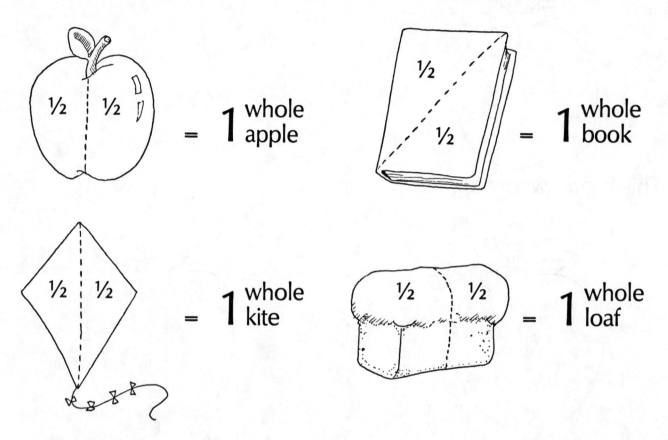

Draw a line to cut each shape in half.

Now draw a half to make each shape whole again.

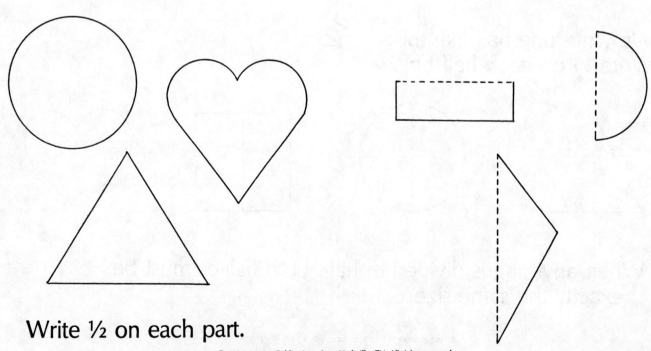

Write ½ on each part.

To make one half, you must cut exactly in the middle, so that both parts are exactly the same size.
If the shape is divided in half, write yes.
If the shape is not divided in half, write no.

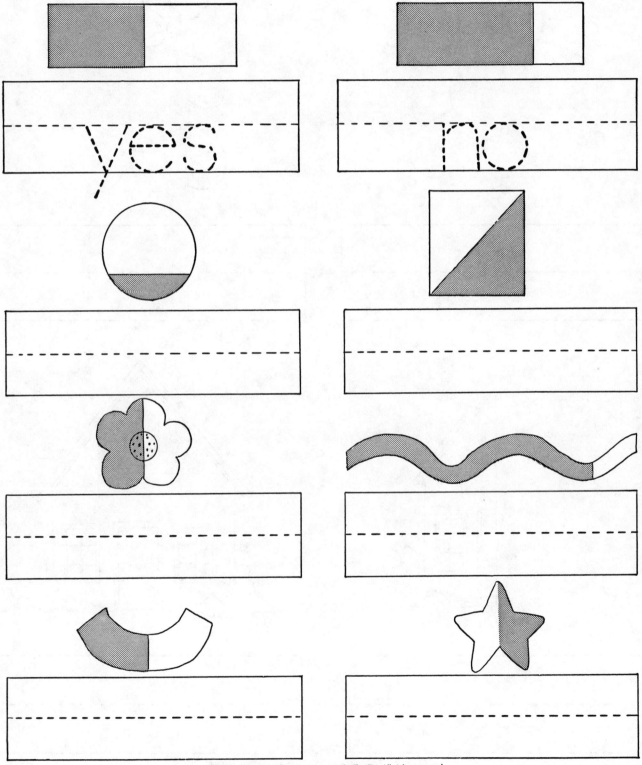

**Write the word <u>half</u> under each picture that shows one half.
Write the word <u>whole</u> under each picture that shows a whole.**

Half 'N Half

Follow the dotted lines to cut the shapes in half.

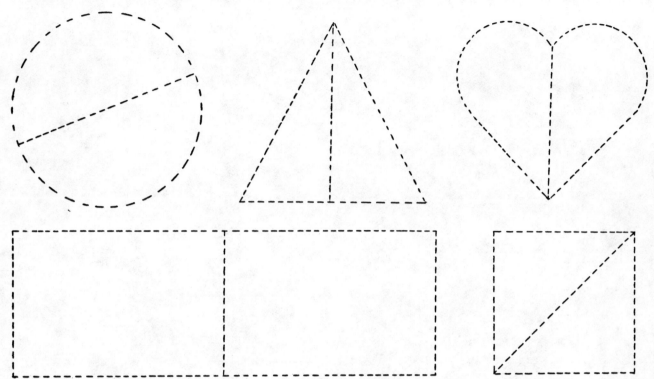

Paste them on the matching shapes below to make them whole again.

 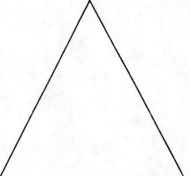 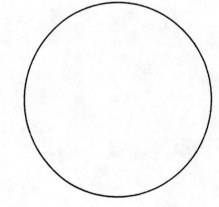

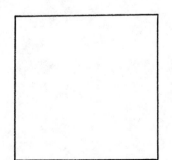

LIVING WITH NUMBERS

Numbers . . . Numbers . . .

Numbers are a very important part of everyday life.
We need and use numbers for many reasons.

Everywhere . . .

Look at all the ways we use numbers. In this picture, circle every number you can find. Tell how or why each one is important.

Special Numbers

Some numbers are special . . . just for you and your family. The pictures below tell about these numbers. Can you guess what they are?

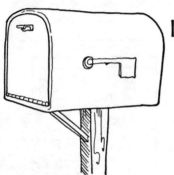

If you live in the country, your address might be on a mailbox like this one.

If you live in an apartment, your apartment number might be on a locked mailbox like these.

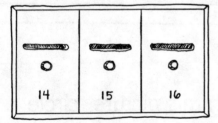

If you live in a big city, your address might be on your house.

Write your address here.

Another special number is one that your friends use to call you on the telephone. Each family or business has its own telephone number. It is not the same as anyone else's telephone number.

Write your telephone number here.

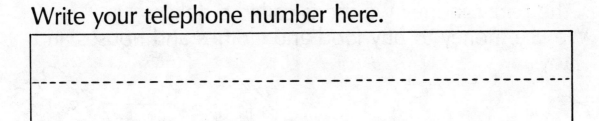

Then make a pretend call to your house by pushing the correct buttons on this telephone.

Make another "fun" call to a good friend whose number you know.

Numbers With Buying Power

Most people of the world use money to buy the things they need.

Around the world, money has many different names, but the paper money we use is called a <u>dollar</u>.

We use money to buy food and clothes and houses and toys.

There are many kinds of dollars. Each kind has different words and pictures, but they all look something like this:

Write the word <u>dollar</u>.

This is the mark or sign that says dollar.
It looks like the letter S with a straight line through it.

Make some dollar signs in this box.

If you bought a toy at the store, it might cost five dollars.
We write five dollars like this: **$5**

$5 is the price of the toy. That is how much you must pay to buy the toy.

Write the price of the toy on its tag.

Write the words for its price.

- -

Each toy on this page has a price tag. It tells how many dollars you will need to buy the toy.
Circle the words that tell the correct price of each toy.

Pretend you are a salesperson in a clothing store.
Your job is to write the correct price on each tag.

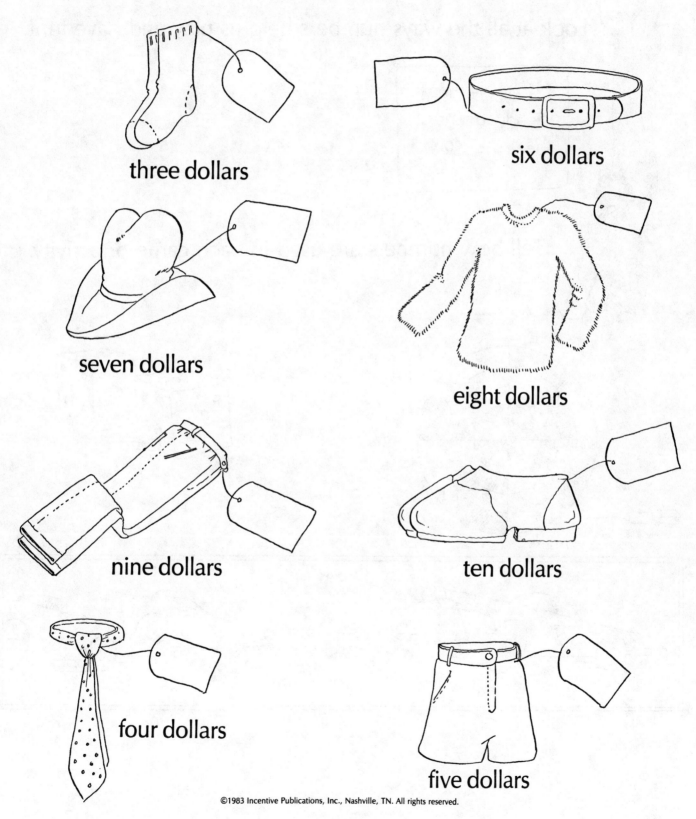

Numbers Are for Fun, Too!

Look at all the ways numbers help us play and have fun!

... 98, 99, 100! Ready or not, here I come!

Tell how numbers are used in each game or activity.

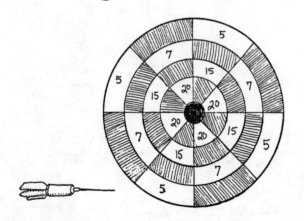

See if you can fill in the numbers to finish this counting rhyme.

one, two

buckle my 🥿

_____

shut the 🚪

_____

pick up 🪵

_____

lay them straight

_____

a big, fat

Numbers To Count Down!

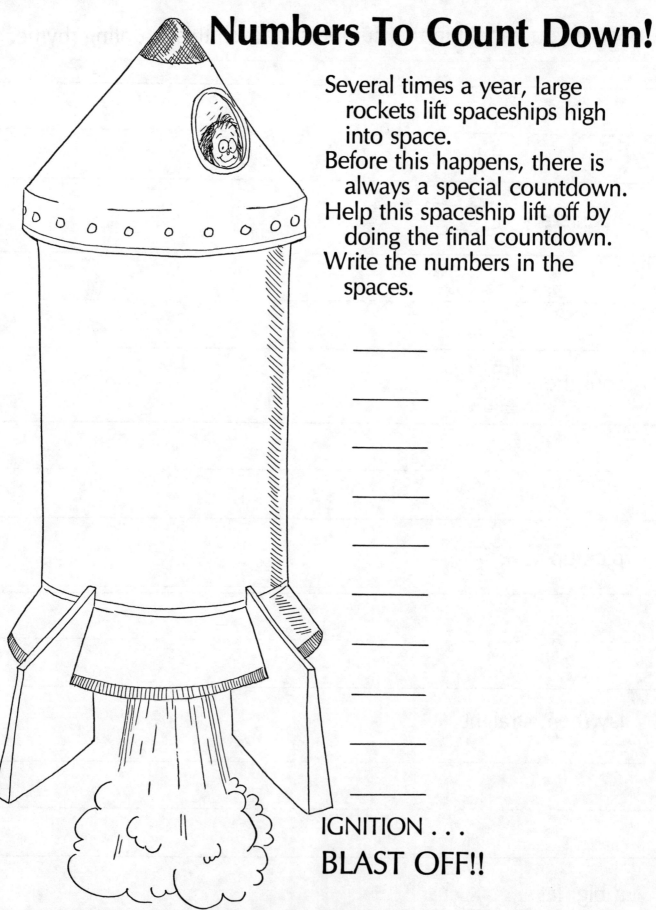

Several times a year, large rockets lift spaceships high into space.
Before this happens, there is always a special countdown.
Help this spaceship lift off by doing the final countdown.
Write the numbers in the spaces.

___
___
___
___
___
___
___
___
___

IGNITION . . .
BLAST OFF!!

These runners are in a marathon race. Can you tell who is winning? Each runner is wearing a number.

Which runner is ahead?
Which runner is farthest behind?
Which two runners are closest together?
Which runner has just passed a tree?
Which runner is nearest the lake?
Which runner do you think will win the race?

©1983 Incentive Publications, Inc., Nashville, TN. All rights reserved.

Numbers To Grow On . . .

Scales tell how heavy something is or how much it weighs.
Some scales measure in pounds.
Each mark on the scale stands for one pound.

This baby weighs eleven pounds.
The arrow points to the numeral 11.

The baby weighs _____ pounds.

Can you tell how much this kitten weighs?

It weighs _____ pounds.

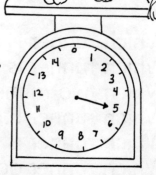

Use a scale to find out how much you weigh.
Write your weight here.

_____

Read the scales to tell how much each object weighs.

_____ pounds _____ pounds _____ pound

Find the objects below. Weigh each on your own scale. Tell how much each weighs.

three books a child a grown-up

_____ pounds _____ pounds _____ pounds

More Numbers To Grow On . . .

To find out the size of things, we measure.
Rulers and measuring sticks tells us how long or tall things
  are.
Their marks tell us how many inches and feet or how
  many centimeters and meters an object is.

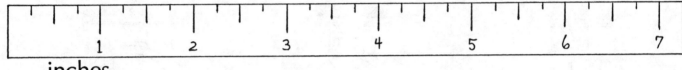
inches

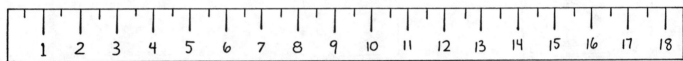
centimeters

Which bookworm is tallest? Color it green.
Put a red cap on the shortest one.

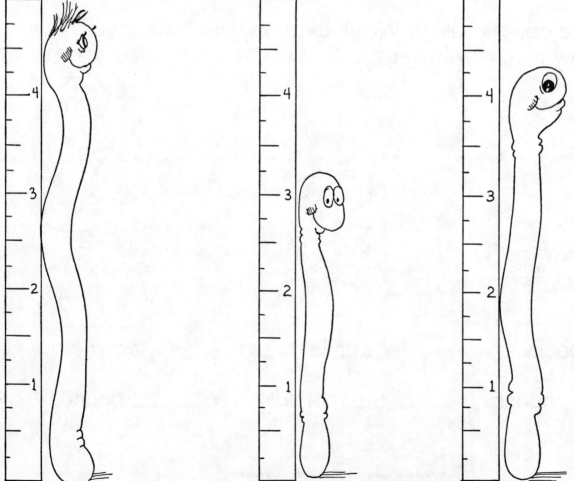

This is a ruler.

To measure with a ruler, you must put the ruler and the object you want to measure side by side, like this.

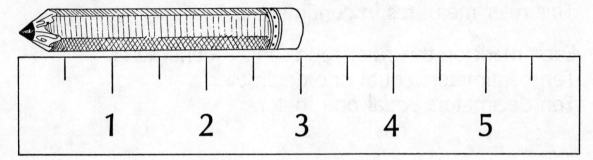

Can you tell how many inches long the pencil is?

If you said three inches, you are exactly right.

Cut carefully on the dotted line to remove the ruler from this page. Use it to measure the objects on the following pages.

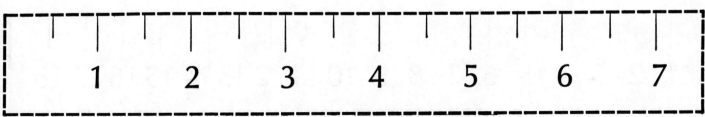

This ruler measures in centimeters.

Each mark on this ruler represents a centimeter.
Ten centimeters equal one decimeter.
Ten decimeters equal one meter.

Can you tell how many centimeters long the toothbrush is?

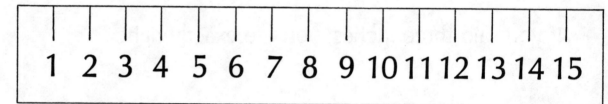

If you said ten centimeters, you are exactly right.

Cut carefully on the dotted line to remove the ruler from this page. Use it to measure the objects on the following pages.

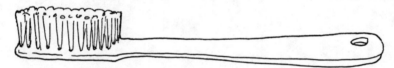

Before there were measuring sticks, people used whatever
   they had handy to measure things.
Usually, they used

fingers and hands   feet   arms

Use your pointer finger to measure the sleepy alligator on
   this page.
How many fingers long is it? _____

Now measure the alligator with your big toe.
How many toes long is it? _____

Try measuring the alligator with a paper clip.
How many paper clips long is it? _____

Use your paper rulers to measure the alligator one more
   time.
How long is it in inches? _____
In centimeters? _____
(If your answer was about six inches or about
   fifteen centimeters, you are very good at measuring!)

Why do you suppose people had to invent measuring sticks?

©1983 Incentive Publications, Inc., Nashville, TN. All rights reserved.

Use your paper rulers to measure the objects on this page.

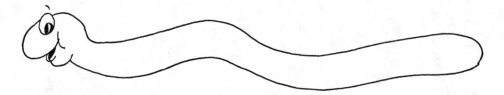

How many inches? _____

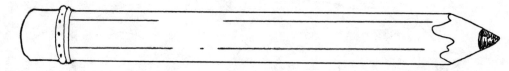

How many centimeters? _____

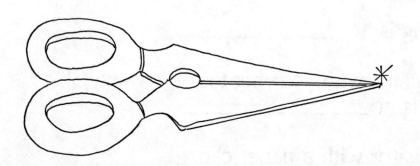

How many inches? _____

How many centimeters? _____

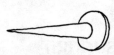

How many inches? _____

Long Numbers

How many centimeters long is the road from your house to your school or your grandmother's house? Could you measure it with a ruler?

That might take a very long time. You would probably have to crawl along on your hands and knees with your ruler for many days.

That is why we use miles or kilometers to measure very long distances.

Maps and road signs tell us how far it is from home to faraway places.

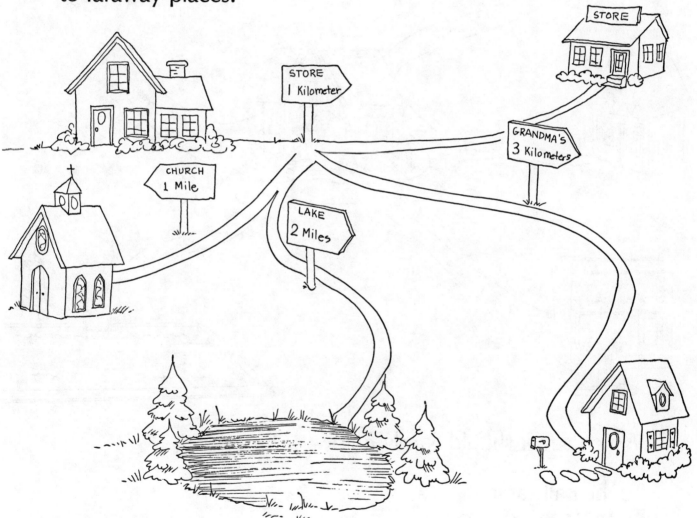

Pretend this is your home.

Read the signs and tell how far you must go to get to each faraway place.

Numbers for Travel

City streets often have number names.
Help Sandy choose the proper streets to get his errands done.

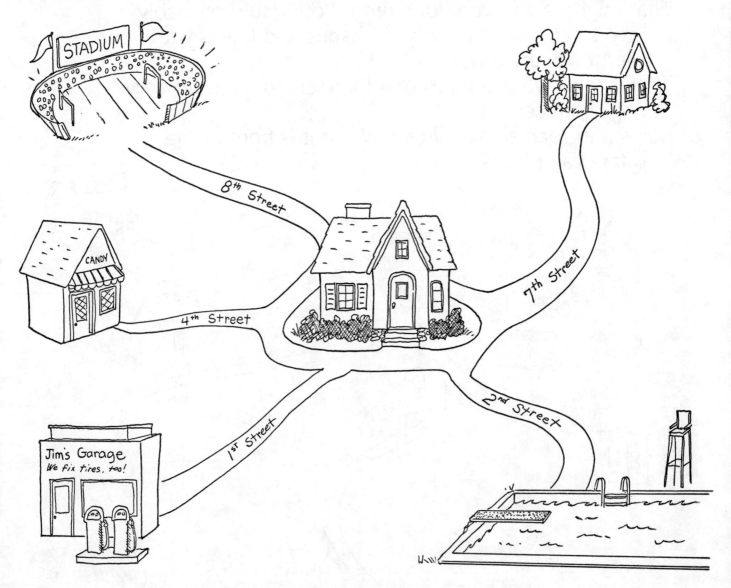

Which street should Sandy follow . . .

To the ball game?
To buy some sweets?
To get a tire fixed?
To visit a friend?
To go swimming?

Very long, important highways or roads usually have numbers instead of names. Maps tell the number of each highway.

Betty gave Donald these directions to get from his house to her house. See if you can help Donald get there.

Dear Donald,
   Take Hwy. 10 to Hwy. 405.
   Stay on Hwy. 405 to Hwy. 101.
   Take Hwy. 101 to Hwy. 75.
   Take Hwy. 75 to my house.

         Love,
         Betty

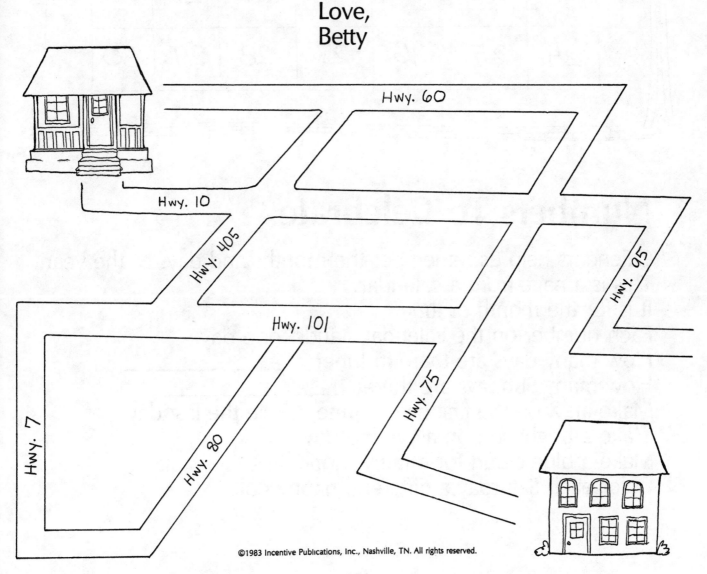

JUNE

Sunday	Monday	Tuesday	Wednesday	Thursday	Friday	Saturday
					1	2
3	4	5	6	7	8	9
10	11	12	13	14	15	16
17	18	19	20	21	22	23
24	25	26	27	28	29	30

Numbers To Celebrate

Calendars help us remember the months and days of the year.
This is a page from a calendar.
It is for the month of June.
Each number on the calendar stands for a day.
How many days are there in June? _____
How many Sundays are there? _____
Make an X on the first day of June. Circle the last day.
Make a bright sun on a Wednesday.
Make a blue cloud for a rainy Monday.
Color each Saturday a different, happy color.

©1983 Incentive Publications, Inc., Nashville, TN. All rights reserved.

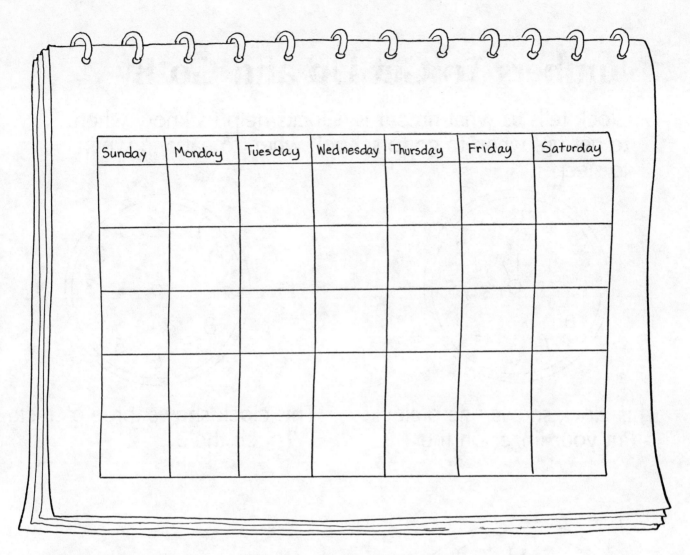

Make the calendar page for your birthday month.
Color the space that shows your birthday a happy color.
What day of the week is your birthday this year? _____
Write the date of your birthday:

_____   _____
name of the month   number of the day

Draw a birthday cake for yourself. On the top of it, put the number of candles you will need for your next birthday.

Numbers To Get Up and Go By

A clock tells us what time it is. Clocks help us know when to get up, when to go to school, when to eat and when to sleep.

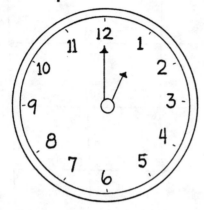

This clock shows one o'clock. Put your finger on the 1.

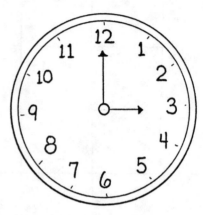

This clock shows three o'clock. Touch the 3.

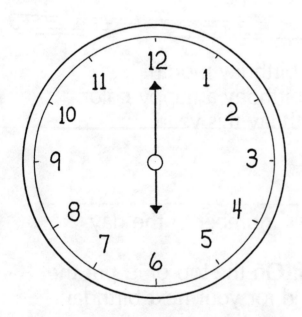

What time does this clock show? If you said six o'clock, you are right.

What do you do about six o'clock in the evening?
What are you usually doing at six o'clock in the morning?

This clock shows ten o'clock.
The short hand points to the number that tells the hour.

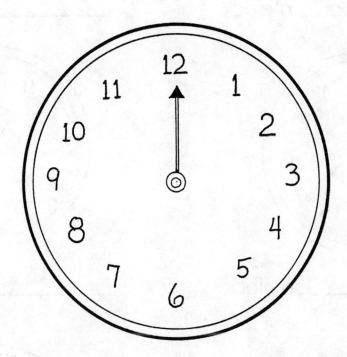

Make the short hand point to any hour you choose.
What time does the clock show?

Write it.

What Time Is It?

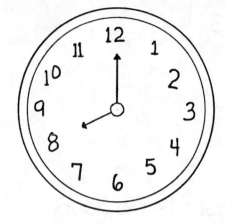

_____ o'clock

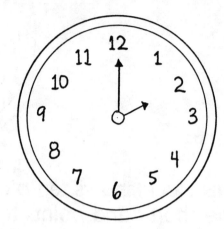

_____ o'clock

_____ o'clock

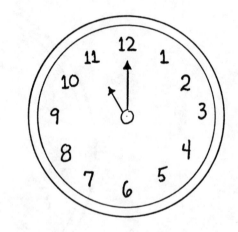

_____ o'clock

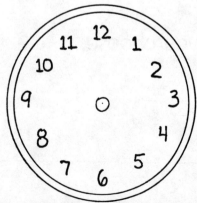

Make this clock show the time to get up in the morning.

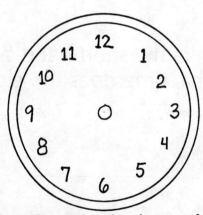

Make this clock show the time for lunch.

This Is Buzzy

The pictures tell you what time Buzzy does things. Can you make the short hour hand on each clock tell the correct time?

Buzzy gets up at seven o'clock.   Buzzy eats breakfast at eight o'clock.

Buzzy goes to school at nine o'clock. Buzzy plays ball at three o'clock.

Buzzy takes a bath at six o'clock.   Buzzy is fast asleep at ten o'clock!

©1983 Incentive Publications, Inc., Nashville, TN. All rights reserved.

Numbers for Finding Things

Pretend that you are a delivery person. You must deliver packages and flowers to several people in a large apartment building.

Use the directory below to find on which floor each person lives.

"Touch" that button on the elevator with the correct color crayon.

Directory

Ms. Orange . . . . 4th floor
Mr. Gray . . . . 7th floor
Mr. Green . . . . 8th floor
Ms. Brown . . . . 6th floor
Ms. Black . . . . 2nd floor
Ms. Blue . . . . 5th floor
Mr. Yellow . . . . 3rd floor
Ms. Red . . . . 10th floor

Numbers To Order and Eat

Numbers make it easy to order from some restaurant menus. Menus tell the cost of each item.

MENU
1. Salad Delight $1.25
2. Soup-errific! $1.00
3. Diet Double $2.00
4. Chicken Lickin' $3.25
5. Hamburger Heaven $2.75
6. Hey, Hot Dog! $1.75
7. Rah! Rah! Ribs $4.50

Tea . . . . $ .75   Milk . . . . $ .60
Coffee . . . . . $ .75   Soda . . . . . . $ .65
Ice Cream . . . . $ .50 a scoop

What item number would you order if . . .

You loved chicken? _____

Your favorite meal was hot dogs? _____

You were on a diet? _____

You were a salad lover? _____

You wanted to spend exactly $1.00? _____

People who are very careful about their health and body fitness often use numbers to help them choose the right foods. They count the calories in each food.

Choose from this list three of your favorite snacks. Write them below and tell how many calories in each.

```
1 peanut butter sandwich .......... 225
1 small banana ..................  81
1 medium apple ..................  80
1 hard-boiled egg ...............  81
1 large dill pickle ..............  15
½ cup red raspberries ...........  41
1 cucumber......................  29
1 chocolate bar ................. 150
1 cup watermelon ................  42
10 pretzel sticks................  12
7 fresh shrimp ..................  70
1 cup cherries ..................  65
1 cup puffed rice ...............  55
5 saltine crackers ..............  60
½ cup cantaloupe ................  24
```

My Favorite Snacks

Snack	Calories
1. _____	_____
2. _____	_____
3. _____	_____

Which snack has the fewest calories?
Which has the most?

Numbers for Cooking and Storing Things

Sometimes we measure things in spoons and cups and bottles.
This cup will measure one cup of water.

one cup

This cup will measure one-half cup of water.

one-half cup

Can you tell how many half cups it takes to fill one cup?
Fill a half cup with water. Pour it into a cup. Fill the half cup again. Pour it into the cup. Is the cup full?
How many half cups did you use to fill it?
Write the numeral.

_____

These are measuring spoons.
¼ teaspoon
½ teaspoon
1 teaspoon
1 tablespoon

Cooks use measuring spoons to measure ingredients. Can you use measuring spoons and follow the recipes to make something good to eat? Try it!

Cinnamon Toast
Mix:
   4 tablespoons sugar
   ½ teaspoon cinnamon
Make toast.
Butter it.
Sprinkle the cinnamon-sugar mixture
   on the toast.

Chocolate Milk
1 cup milk
2 tablespoons chocolate syrup
Place both in a glass. Stir with a
   spoon.

Sometimes we need bigger things to measure liquid.
Then we use quarts or gallons or liters.

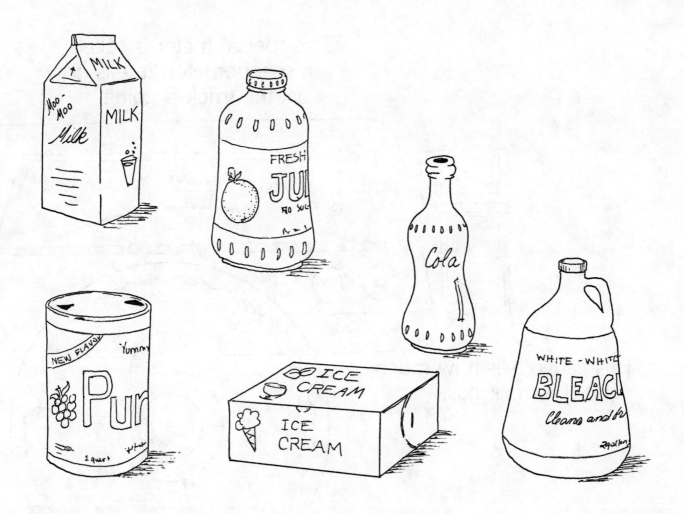

Here are some things that come in quarts and gallons and liters.
Can you tell what each one is and where you might find a real one like it?

Great Going With Gauges and Meters

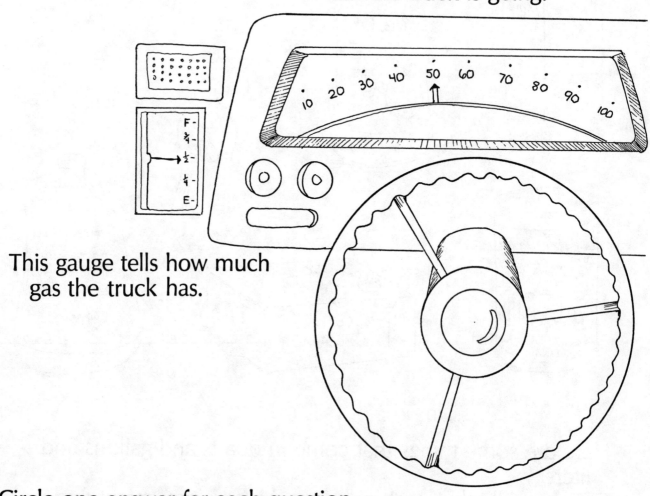

This special meter is called a speedometer. It tells how fast the truck is going.

This gauge tells how much gas the truck has.

Circle one answer for each question.

How much gas does the truck have?   ½   ¼   ¾

How fast is the truck going?   20   45   50

Numbers Hot and Cold

A thermometer has numbers that tell us how hot or cold something is. We say it tells us the temperature.

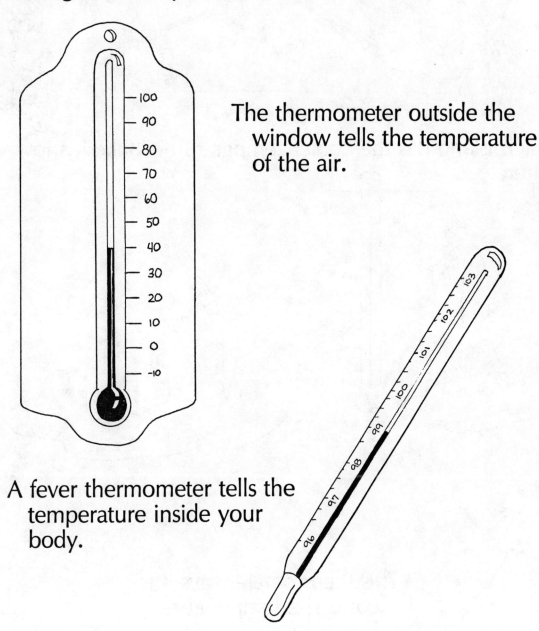

The thermometer outside the window tells the temperature of the air.

A fever thermometer tells the temperature inside your body.

Can you read these thermometers?

Which patient has the warmer body? Draw an ice pack on her head.

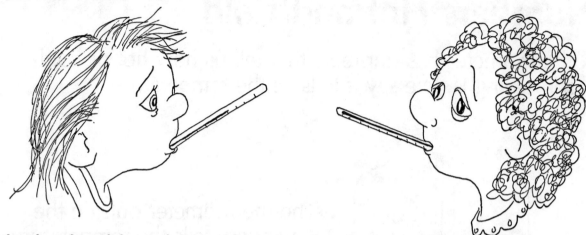

Which island has the cooler temperature? Make it snow there.

This thermometer says it is cold. The temperature is _____.

Numbers Star on the Big Screen

This is a computer.
Numbers are important to computers.
There are hundreds of kinds of computers. Most of them need numbers to help them do their work.
This computer is programmed to tell about you. Make it tell your story.

Name _____

Address _____

_____

Height _____ Weight _____

Age _____ Date of birth _____

Number of brothers _____ Number of sisters _____

Phone number _____ Number of pets _____

Number of teeth lost so far _____

Computers can teach school, take pictures, forecast the weather, run cars and trains, make music and cartoons, play games and control rockets and robots. Can you tell some other things computers can do?

TEST YOURSELF

X Marks the Answer

Which is shorter?

Which is shortest?

Which is larger?

Which has the fewest feet?

$4 or $7   Which is greater?

    Which is a pair?

Circle the Set

Circle the set of four.

 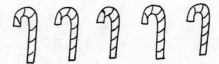

Which set is greater?

Find the sets that are the same.

   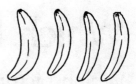

Find the set with the fewest.

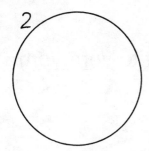 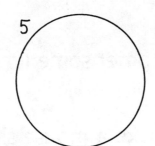 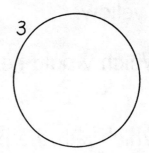

Draw a correct set in each circle.

©1983 Incentive Publications, Inc., Nashville, TN. All rights reserved.

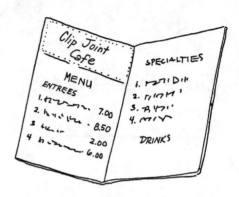
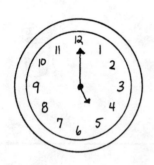
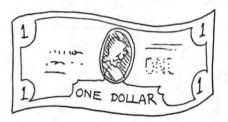

Which Is Which?

Which picture shows something that measures hot and cold? Draw a squiggly line around it.

Which will tell you the day of the week? Color it yellow.

Which would help you order some lunch? Write <u>yum yum</u> by it.

Which tells the hour? Make a mouse sitting on it.

Which can you use to buy something? Write its name under the picture.

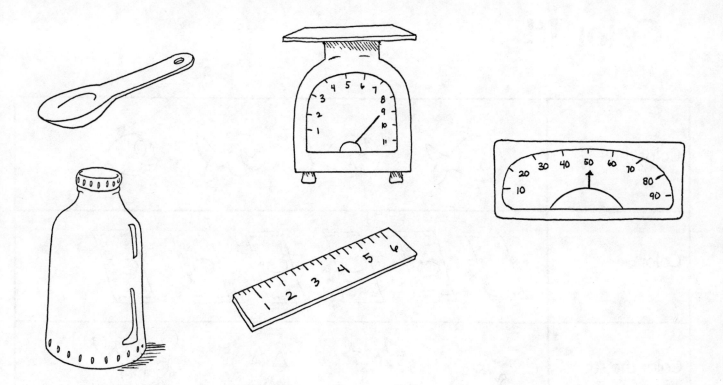

Which measures the cinnamon you will need to make toast?
Sprinkle it with brown.

Which stores liquid for drinks?
Fill it with delicious, yellow lemonade.

Which two things would help you measure something?
Draw something heavy on one.
Draw something long by the other.

Which would tell you how fast something was moving?
Draw something that can move fast near it.

Color It!

Color three.	
Color $4.	
Color the set of six.	
Color the things that measure.	
Color the sign that shows the most miles.	
Color the halves.	
Color five o'clock.	

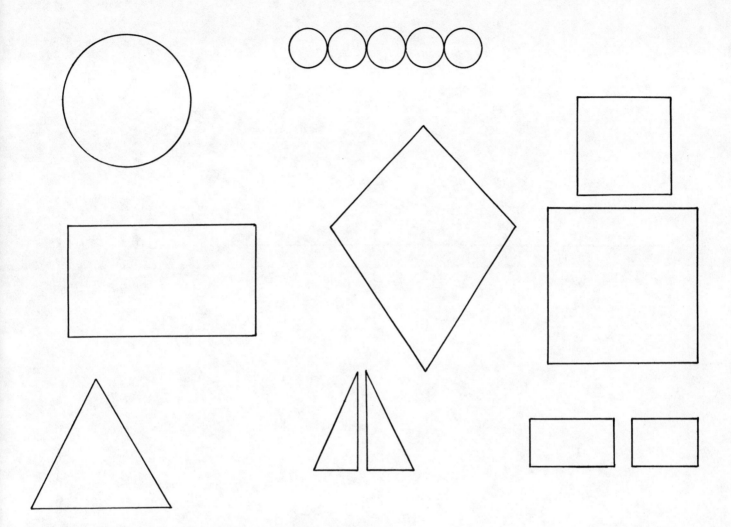

Quick-Change Artist

1. Add lines to the rectangle to make it a toy.
2. Make the set of two triangles into a boat.
3. Make the diamond shape become part of a kite.
4. Use the set of five circles to make an animal.
5. Make a robot using the set of two squares.
6. Connect the set of two rectangles to make any object you like.
7. Use the circle to make a funny face.
8. What's left? Use it to make a home for an American Indian.